Harold the Super Pet

by Sally Rippin

illustrated by Simon Scales

EDUCATORS PUBLISHING SERVICE
Cambridge and Toronto

Series Authors: Kay Kovalevs and Alison Dewsbury
Series Consultant: Melinda Rice
Commissioning Editors: Sarah Russell, Bonnie Lass, and Sethany Rancier Alongi
Senior Editorial Manager: Sheila Neylon
Text by Sally Rippin
Illustrated by Simon Scales
Edited by Simone Calderwood, Laura A. Woollett, Stacey Nichols Kim, and Bonnie Lass
Designed by Kathryn de Reus and Maria Vallianos

Making Connections program developed by Educators Publishing Service, a division of School Specialty Publishing and by Harcourt Education, a division of Reed International Books Australia Pty Ltd

ISBN 0-8388-3316-0

2010 2009 2008 2007 2006
10 9 8 7 6 5 4 3 2 1

Printed in China

Max wanted a dog. All his friends had dogs.

"Could we get a dog, Dad?" Max asked.

"Our apartment is too small for a dog," said Dad. "A dog needs to have a yard to run in."

“Could we get a small dog then? Bruno has a small dog that can stand up on two legs,” Max said.

“Even small dogs need a yard, Max,” Dad said. “We don’t have one. What about another type of animal? How about a goldfish?”

Max frowned. “Dad! You can’t teach a goldfish tricks!”

The next day, when Dad came home from work, he had a big box in his hands and a big smile on his face.

"I have a present for you," Dad said to Max. "I've brought you a pet."

"Is it a dog?" asked Max, jumping up and down.

"No," said Dad. "It's not a dog, Max. Why don't you open the box and see?"

Max opened up the box. Inside was a cage. Inside the cage was a tiny ball of brown fur.

"What is it?" asked Max. He poked gently at the ball.

"It's a hamster," said Dad.

"Hmmm . . ." said Max. "He's very small. Can he do any tricks?"

Carefully, Max picked up the tiny ball of fur. Out popped a pink nose and two bright eyes.

"Of course he can do tricks!" said Dad. "His name is Harold. He's a Super Pet!"

Max frowned. This pet didn't look super. In fact, he didn't even look like much of a pet at all.

“Remember,” said Dad, “if you have a pet, you have to take care of it.”

“Yes, Dad,” said Max.

“You have to feed it, play with it, and make sure it has fresh water every day.”

“Yes, Dad,” said Max.

"And you have to make sure Harold's cage is shut tight, especially at night," said Dad. "Hamsters are very active at night."

"Yes, Dad," said Max.

Max lifted Harold up to his face. The hamster opened a sleepy eye.

"You don't look very active to me," whispered Max as he put Harold back into his cage.

That night, when Max was nearly asleep, he heard a strange hum coming from his closet. It sounded just like his remote-controlled car.

Max sat up and the noise stopped. He got out of bed and crept across his room to the closet. Max threw open the door.

Max looked down. The car was sitting right where he had left it, but it was switched on and it was buzzing. He turned it off.

As Max tiptoed back to bed, he peered around his moonlit room. He saw Harold's cage. Max checked to make sure the cage door was shut. It was. Harold was safely inside, still asleep.

"Active at night," Max said to himself. "Yeah, right!"

He wished his dad had gotten him a dog. With a sigh, he got back into bed and went to sleep.

The next night, just as Max was falling asleep, he heard the same noise again. This time, he sat up quickly. Out of the corner of his eye, he saw his remote-controlled car speed across the room!

Max jumped out of bed, but the car had already stopped just in front of Harold's cage.

"Harold?" Max said, puzzled. But Harold was sleeping quietly in the corner of his cage.

The third night, Max hid a flashlight under his pillow. When the noise began, Max woke suddenly. He sat up quickly, shining the flashlight at Harold's cage.

Just then, the car whizzed past and Max saw Harold sitting inside! His fur was flying back in the air and he was holding on to the steering wheel with all his might.

"Harold!" Max cried, jumping out of bed. "I can't believe it! Is that you? Are you driving a car?"

The car stopped, and Max ran to pick up the tiny hamster before he could scamper back into his cage.

"Dad! Dad!" Max shouted, running down the hallway to his father's bedroom. "You won't believe what Harold can do! You were right. He really is a Super Pet!"

Max's dad came out of his room wearing his bathrobe and smiling a secret smile. Hidden in his bathrobe pocket was a square metal box, with a long antenna sticking out of it. Can you guess what it might have been?